Department of Homeland Security
Office of Inspector General

Management Advisory Report
On Cybersecurity

Council of the
INSPECTORS GENERAL
on INTEGRITY and EFFICIENCY

This report was prepared on behalf of Council of the Inspectors General on Integrity and Efficiency

OIG-11-121   September 2011

September 30, 2011

Preface

At the request of the Council of the Inspectors General on Integrity and Efficiency (CIGIE) Homeland Security Roundtable (HSR) and with the approval of the CIGIE Executive Council, the Department of Homeland Security (DHS) Office of Inspector General (OIG) chaired a Working Group of attorneys and information technology (IT) professionals (IT security professionals, IT auditors, and other IT practitioners) and other cybersecurity experts from OIGs of various sizes, including representatives of the presidentially appointed and designated federal entity Inspectors General (IG) community.

The CIGIE Cybersecurity Working Group was charged with undertaking a two-part review in which it would (1) identify recommended practices for maintaining the integrity of OIG IT systems and protecting them against internal threats and vulnerabilities and (2) examine the role of the IG community in current federal cybersecurity initiatives.

I am pleased to provide the CIGIE Cybersecurity Working Group's recommended practices for maintaining the integrity of OIG IT systems. This report is the product of the first part of the review. It is based on the subject matter expertise of IT specialists from a representative group of the IG community, discussions with industry professionals, legal research, and a review of applicable websites and documents. These recommended practices are intended to help the IG community address the many issues and demands that OIGs and government managers face today. The recommendations herein have been developed to the best knowledge available to the Working Group. We trust that this report will result in more secure OIG IT systems. DHS OIG would like to express its appreciation for the considerable amount of time dedicated to this effort.

I would like to acknowledge the support provided to this cybersecurity effort by all the working group participants listed in appendix C. Of particular note is the work of Chris Orcutt, Patrick Nadon, Jefferson Gilkeson, Jaime Vargas, Rachel Magnus, Phyllis Bryan, Adam Berlin, Robert Duffy, and Rene Lee to produce a final document that represents the needs of the IG community.

Charles K. Edwards
Acting Inspector General

Table of Contents

Abbreviations

CIGIE	Council of the Inspectors General on Integrity and Efficiency
CIO	Chief Information Officer
CIS	Center for Internet Security
CM	configuration management
CNCI	Comprehensive National Cybersecurity Initiative
COOP	Continuity of Operations Plan
CPR	Cyberspace Policy Review
DHS	Department of Homeland Security
DOD	Department of Defense
EH-11	Eagle Horizon 2011 Exercise
FDCC	Federal Desktop Core Configuration
FedRAMP	Federal Risk and Authorization Management Program
FEMA	Federal Emergency Management Agency
FICAM	Federal Identity, Credential, and Access Management
FIPS	Federal Information Processing Standards
FISMA	*Federal Information Security Management Act*
GPEA	*Government Paperwork Elimination Act*
GSA	General Services Administration
HIPAA	*Health Insurance Portability and Accountability Act of 1996*
HSPD	Homeland Security Presidential Directive
HSR	Homeland Security Roundtable
ICAM	Identity, Credential, and Access Management
IG	Inspectors General

ITD	Information Technology Division
IT	information technology
MC	Mission Critical
MEF	Mission Essential Function
NIST	National Institute of Standards and Technology
NPE	non-person entity
OIG	Office of Inspector General
OMB	Office of Management and Budget
PIV	Personal Identity Verification
PKI	Public Key Infrastructure
SANS	System Administration Networking and Security
SCAP	Security Content Automation Protocol
SCM	security configuration management
SP	Special Publication
STIG	Security Technical Implementation Guides
TIC	Trusted Internet Connection
VPN	Virtual Private Network

Council of the
INSPECTORS GENERAL
on INTEGRITY *and* EFFICIENCY

The CIGIE was statutorily established as an independent entity within the executive branch by the *Inspector General Reform Act of 2008*, Public Law 110-409. The mission of the CIGIE is to—

- Address integrity, economy, and effectiveness issues that transcend individual government agencies; and

- Increase the professionalism and effectiveness of personnel by developing policies, standards, and approaches to aid in the establishment of a well-trained and highly skilled workforce in the federal IG community.

Membership

- All IGs whose offices are established under either section 2 or section 8G of the *Inspector General Act*, or pursuant to other statutory authority (e.g., the Special IGs for Iraq Reconstruction, Afghanistan Reconstruction, and Troubled Asset Relief Program)

- The IGs of the Office of the Director of National Intelligence (or at the time of appointment, the IG of the Intelligence Community) and the Central Intelligence Agency

- The IGs of the Government Printing Office, the Library of Congress, the Capitol Police, the Government Accountability Office, and the Architect of the Capitol

- The Controller of the Office of Federal Financial Management

- A senior-level official of the Federal Bureau of Investigation, designated by the Director of the Federal Bureau of Investigation

- The Director of the Office of Government Ethics

- The Special Counsel of the Office of Special Counsel

- The Deputy Director of the Office of Personnel Management

- The Deputy Director for Management of the Office of Management and Budget (OMB)

<u>CIGIE HSR</u>

Since September 11, 2001, protecting our Nation has been a paramount concern of the entire federal establishment. The IG community plays a significant role in reviewing the performance of agency programs and operations that affect homeland security. To a large extent, this has been accomplished through collaborative efforts among multiple OIGs.

On June 7, 2005, the President's Council on Integrity and Efficiency Vice-Chair established a President's Council on Integrity and Efficiency HSR. The roundtable supports the IG community by sharing information, identifying best practices, and participating on an ad hoc basis with various external organizations and government entities. The CIGIE Cybersecurity Working Group was formed under the auspices of the HSR.

Executive Summary

Computers, the Internet, and other electronic assets have become integral to the effective functioning of the federal government, its programs, and daily public life. These assets have become the targets of people with malicious intent, and thus represent an area of increased risk and vulnerability to the federal government. The community of Inspectors General must be proactive in preventing and addressing issues relating to cybersecurity, both in its oversight capacity and in its operational role. To that end, the Council of the Inspectors General on Integrity and Efficiency Cybersecurity Working Group was charged with identifying measures that the Inspector General community can take to protect itself against cyber attacks.

This report covers four areas identified as cybersecurity challenges facing the Inspectors General community: (1) asset management and leveraging resources; (2) identity, credential, and access management; (3) incident detection and handling; and (4) scalable trustworthy systems. The topics are not exhaustive of all cybersecurity issues. They were identified by the Cybersecurity Working Group, using the DHS *Roadmap for Cybersecurity Research*, as the most salient and relevant issues facing Council of the Inspectors General on Integrity and Efficiency community members.

The report offers recommended practices for the Inspectors General community taking into consideration the different risks or vulnerabilities of each OIG based on the degree to which information technology systems are dependent upon or connected to their parent agencies and whether they have sufficient human and financial resources to secure their information technology systems effectively. Although each element of the report will not apply to each unique OIG/parent agency structure, the report provides a foundation for understanding some of the most salient issues facing our organizations and the solutions to these issues. Generally, the cybersecurity issues faced by the Inspector General community are the same as those faced government-wide; however, for each office, the mission, type of information collected, and the type of work may impact the relative priority of the problems and issues. A subsequent report will address the Inspectors General community's role in federal cybersecurity initiatives.

Background

IT has become pervasive in every way, from our phones and other small devices to our enterprise networks and the infrastructure that runs our economy. As the critical infrastructures of the United States have become increasingly dependent on public and private IT networks, the potential for widespread national impact resulting from disruption or failure of these networks has also increased. This report presents key topics and recommendations that the IG community can consider and use when securing existing systems and adopting new technologies.

Cybersecurity is a broad and complex area of study. A six-month review cannot fully address all of the topics in the cybersecurity arena. The Working Group focused its efforts on the four challenges that are most salient to improving cybersecurity in the IG community: (1) asset management and leveraging resources; (2) identity, credential, and access management; (3) incident detection and handling; and (4) scalable trustworthy systems.

As part of this effort, the Working Group surveyed the IG community to gather information on current and planned initiatives to address these challenges. It received responses from 41 of 79 members of CIGIE. The survey results, which are compiled and summarized in appendix B of this report, were used to analyze cybersecurity initiatives and trends in the IG community.

Results of Review

The Working Group identified four areas which, if properly understood, designed, and monitored, provide the IG community and respective agencies with assurance that risks associated with the areas are minimized. The report is organized to reflect the logical steps taken to secure an infrastructure: identifying and managing assets, controlling and monitoring access to those assets, and managing detection and handling incidents. Last, the Working Group analyzed how emerging technology such as cloud computing can be leveraged into a trustworthy system.

Asset Management and Leveraging Resources

Asset management is the set of organizational practices that identify and control all elements of hardware and software in an organization. It is the first step toward managing network weaknesses, device vulnerabilities, and configuration challenges. Computer systems and the information they store are critical assets that support an organization's mission. Protecting critical assets from cyber threats is an essential management function, and therefore, an understanding of asset management and its processes is fundamental to the IG community organizations.

Asset management refers to the tracking of all tools, their accessories, and what each tool needs in order to perform as intended. Hardware asset management is the management of physical components, while software asset management focuses on software assets, which include installation tracking for licensing, versioning, and upgrades. Asset management may be compared to the organization of a tool-shed. Tools are typically owned for use as needed; asset management best practices can help to find tools that cannot be located immediately. To do so, users need a list that includes each tool, its location, performance requirements, most recent maintenance by date and type, and when the next maintenance should be performed. Resources and time need to be devoted to keeping all this information updated.

Many processes outside of the cybersecurity function play important roles in cybersecurity asset management. For example, before you can check a computer for necessary security patches, you need to know if the computer exists, its location, and what operating system is installed on it. There also needs to be a defined organizational process for prioritizing, testing, and installing software patches.

From a security perspective, strong asset management can help network administrators identify and manage network weaknesses and device vulnerabilities. Unauthorized and undocumented network-attached devices can leave an organization vulnerable to cyber threats, and unmitigated software vulnerabilities may also leave an organization's networks susceptible to cyber attacks. Proper asset management can also assist with identifying lost equipment and illicit configuration changes when providing incident response support. Finally, cybersecurity asset management is necessary to meet federal guidelines and directives such as OMB Memorandum 10-15, which requires agencies to upload IT inventory information, and OMB Circular A-130, which establishes policy for the management of federal information resources government-wide.

Several asset management best practices have security implications—

- Request and approval process,
- Procurement management,
- Configuration management (CM),
- Vulnerability management, and
- Disposal management.

Below is an overview of each asset management best practice.

Request and Approval Process

The request and approval process is a structured and predetermined series of events that allows for streamlined acquisitions using a standard procedure. In the past, when an IT asset was requested, management would either agree or disagree. Current federal regulations require that work, including security planning, be performed before IT purchases are made, and an approval review is typically performed before assets are purchased.

A standardized request and approval process review ensures that the purchase is necessary, is a good investment, and fits with the current security configuration. The review also ensures that any potential new security risks are reviewed and accepted. This review and approval process should include board members from an organization's financial, functional, IT, and security areas.

Procurement Management

Procurement management defines the processes used to determine which assets best meet the organization's needs. For organization-wide purchases, procurement decisions should be made by a team representing different functional areas of the organization. Soliciting input from people with different organizational and functional expertise helps ensure effective procurement decisions.

This procurement team might include an executive officer, IT asset manager, IT manager, IT technical specialist, functional manager, functional end-user, helpdesk manager, procurement attorney, and security specialist. The team's first meeting should review the inventory of assets to provide information on what is currently in use and what is working well, and to report any recurring asset problems. The team builds a business case for each prospective asset purchase, making a value proposition to support the acquisition decision. This team should establish guidelines for standard asset acquisitions and can consider requests for nonstandard assets. This team should be familiar with current procurement regulations as well as security implications.[1]

---

[1] See 48 C.F.R. pts. 1-99 - Federal Acquisition Regulation. Among other things, parts 39 and 52 provide contract language directing a contractor's computer security responsibilities (see, e.g., 48 C.F.R. §§39.105, 39.107, and 52.239-1).

Configuration Management

CM can be defined as establishing and controlling changes made to hardware and software throughout the life cycle of an information system. CM for security, referred to as security configuration management (SCM), manages and controls security configuration items for an information system. The goal of SCM is to enable security configuration items to reduce risk.

Several different entities publish security baselines for various IT products, including the U.S. Government Configuration Baseline, Defense Information Systems Agency, National Security Agency, and the Center for Internet Security (CIS). These security baselines provide configuration settings to "lock down" information systems and software that might otherwise be vulnerable to attack.

The National Institute of Standards and Technology (NIST) provides guidance to implement SCM in organizations. NIST Special Publication (SP) 800-53, Revision 3, has a family of CM security controls (CM-1 through CM-9). NIST SP 800-128, *Guide for Security-Focused Configuration Management of Information Systems*, and SP 800-53A, Revision 1, *Guide for Assessing the Security Controls in Federal Information Systems and Organizations, Building Effective Security Assessment Plans*, provide guidance on implementing SCM controls. Specifically, NIST SP 800-128 identifies the major phases of SCM and describes the process of applying SCM practices for information systems, including (1) planning SCM activities for the organization, (2) identifying and implementing SCM, (3) controlling and maintaining the configuration of the information system in a secure state, and (4) monitoring the configuration of the information system to ensure that the configuration is not inadvertently altered from its approved state.

The NIST SP 800-53, CM-6 security control requires that agencies establish and document configuration settings for IT products. The Defense Information Systems Agency, a component of the Department of Defense (DOD), has defined baselines called Security Technical Implementation Guides (STIGs) to lock down information systems and software that might otherwise be vulnerable to attack. The STIG website contains links to numerous security baselines for operating systems, applications, and telecommunication equipment. STIGs can assist agencies in producing security baselines for the products they use.

Vulnerability Management

Vulnerability management is the practice of identifying, classifying, remediating, and mitigating vulnerabilities. This practice generally refers to software vulnerabilities in computing systems. However, as with CM, definitions vary in the IT industry, and the lines between CM and vulnerability management tend to blur. Managing a baseline of security configuration items can assist with protecting against vulnerabilities. Mitigating existing vulnerabilities enhances an organization's security configuration baseline.

Vulnerability management is achieved by performing vulnerability assessments. Assessments are typically performed according to the following steps—

1. Cataloging assets and capabilities (resources) in a system,

2. Assigning quantifiable value (or at least rank order) and importance to those resources,

3. Identifying the vulnerabilities or potential threats to each resource, and

4. Mitigating or eliminating the most serious vulnerabilities for the most valuable resources.

NIST SP 800-53 also has a risk assessment family of security controls. These controls require that organizations identify and report vulnerabilities. Vulnerabilities need to be analyzed and their potential impact measured. Vulnerabilities should be remediated, mitigated through compensating controls, or documented with the potential risk to the organization accepted.

Disposal Management

As part of the asset disposal process, organizations need controls to assess and, when appropriate, sanitize sensitive information on assets approved for disposal. For example, computer printers, copy machines, and fax machines may contain sensitive residual information which, if released, could have an adverse effect on the organization or individuals whose personal information is stored on agency assets.

Disposal management also includes transitioning old systems to new ones. The replacement process must be planned carefully to prevent vital business data from being lost or compromised. If the asset being retired is business data, National Archives and Records Administration regulations may apply.[2]

NIST provides several recommendations and guidelines concerning the secure disposal of IT media and equipment. The NIST SP 800-53, Software Integrity family of security controls instructs organizations to handle and retain both information within and output from information systems in accordance with applicable federal laws, executive orders, directives, policies, regulations, standards, and operation requirements.

The NIST SP 800-53, Media Protection family of security controls discusses sanitation requirements of digital and nondigital information system media prior to disposal. NIST SP 800-88, *Guidelines for Media Sanitation*, includes a list of common media types and recommends destruction procedures.

Leveraging Resources

Managing assets can be complex and time-consuming. Therefore, using specialized software to automate the process can be beneficial for (1) assessing and managing organization-wide inventories of hardware and software, (2) ensuring compliance with software licenses and other regulatory requirements, and (3) adding value to the disposal process.

Specifically, automated discovery of hardware and software identifies what systems are connected to the organization's network and where they are located. Moreover, an automated software inventory provides an accurate audit of all software applications installed on client computers across the network. An asset management solution can audit this information quickly and then help an organization separate primary applications from operating system and shareware software. It can identify installations of products for license compliance. It can also identify products no longer in use as well as redundant software, which can result in significant cost savings in licensing and maintenance.

Furthermore, an automated software inventory tells an agency how equipment is configured and when changes are made. It can also look for software downloaded from the Internet, which can threaten the security and integrity of the network. Identifying such software is increasingly important as the number of Trojan viruses increases. Automated vulnerability programs can detect and report on configuration issues, software weaknesses, and missing security patches, all of which can be exploited to gain access to secure networks and computers. Finally, when

---

[2] See, e.g., 36 C.F.R. part 1236 for regulations detailing records management requirements for electronic information.

hardware and software is retired, the inventory can verify that all affected assets have been removed from the network.

According to the *OMB Fiscal Year 2010 Report to Congress on the Implementation of The Federal Information Security Management Act (FISMA) of 2002*, the ideal goal of IT asset management capability is to have 100% of agency assets under an automated asset management system that captures data about each asset and can provide that data within a short period of time. Many solutions exist. NIST maintains a list of Security Content Automation Protocol (SCAP) validated products at http://nvd.nist.gov/scapproducts.cfm.

Ultimately, responsibility for asset management in an organization lies with its senior management. It is up to senior management to promote the organization's computer security program and ensure that the proper resources are available.

Recommendations

We recommend that OIGs consider implementing the following practices, when applicable:

Recommendation #1: As a cost efficiency measure, create baseline software assets to support working capital fund requirements and leverage managed and shared services when available.

Recommendation #2: Consider creating an IT purchasing team of selected individuals from various areas of the organization to ensure that purchases best meet the organization's needs and security requirements.

Identity, Credential, and Access Management

The Federal Identity, Credential, and Access Management (ICAM) Initiative[3] efforts, including those of the IG community, are a key enabler for addressing the Nation's cybersecurity challenges. In recent years, increasing emphasis has also been placed on improving the physical security of the hundreds of thousands of facilities that the federal government owns and leases. In addition to complex physical and logical cybersecurity threats, the federal government faces significant challenges in carrying out its IT capabilities to enable a level of assurance and electronic service delivery (see figure 1).

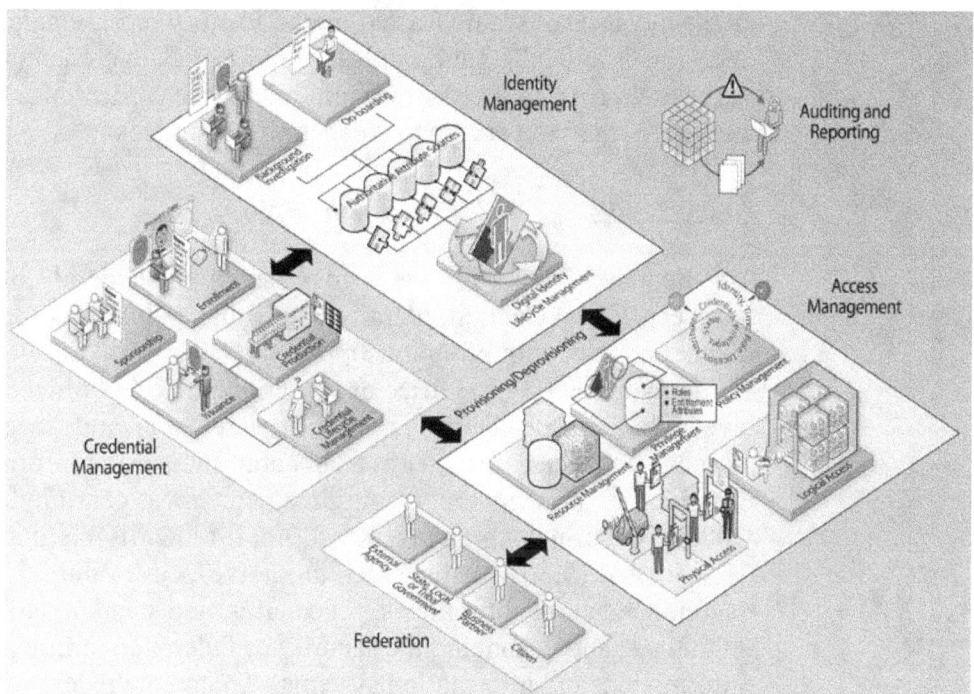

Figure 1: Federal Identity, Credential, and Access Management Roadmap and Implementation Guidance, Version 1.0, November 20, 2009. Figure copyrighted © 2011, Deloitte Development LLC. All rights reserved. Member of Deloitte Touche Tohmatsu Limited.

These challenges lie in the ability to verify the identity of an individual or non-person entity (NPE) in the digital realm and to establish trust in the use of that identity in conducting business.[4] As a result, strong and reliable ICAM capabilities across the entire federal government are a

---

[3] IDManagement.gov is a one-stop shop for citizens, businesses, and government entities interested in identity management activities, including topics related to Homeland Security Presidential Directive 12 (HSPD-12); Federal Public Key Infrastructure; Identity, Credential, and Access Management; and Acquisitions.

[4] NPE is an entity with a digital identity that acts in cyberspace, but is not a human actor. This can include organizations, hardware devices, software applications, and information artifacts.

critical factor in the success of all mission work. A common, standardized, trusted basis for digital identity and access management is needed to provide a consistent approach to deploying and managing appropriate identity assurance, credentialing, and access control services. The approach must also promulgate implementation guidance and best practices, build consensus through government-wide collaboration, and modernize business processes to reduce agency costs for administering and duplicating identity management. Appendix A presents a sample of general laws, regulations, and policies that affect and, in many cases, have initiated today's ICAM programs.

An ICAM plan integrates programs, processes, technologies, and personnel used to create trusted digital identity representations of individuals and NPEs and binds those identities to credentials that may serve as a proxy in access transactions. Those credentials are then used to provide authorized access to an agency's resources.

Governance

The Federal ICAM Initiative is governed by the Federal Chief Information Officers (CIOs) Council, Identity Credential and Access Management Subcommittee, with program support by the General Services Administration (GSA) Office of Government-wide Policy and direct oversight from the OMB. The Identity Credential and Access Management Subcommittee is a subcommittee of the Information Security and Identity Management Committee, which was chartered in December 2008 as the principal interagency forum for identifying high-priority security and identity management initiatives, developing recommendations for policies, procedures, and standards to address initiatives, and enhancing the security of federal government networks, information, and information systems. Today, the federal government is strongly interested in unifying these areas and other identity management initiatives to create a comprehensive and integrated approach to ICAM.

Identity Management

The National Science and Technology Council Subcommittee on Biometrics and Identity Management defines identity management as the combination of technical systems, rules, and procedures that define the ownership, utilization, and safeguarding of personal identity information.[5] The primary goal of identity management is to establish a trustworthy process for assigning attributes to a digital identity and to connect that identity to an individual. Identity management includes the processes for maintaining and protecting the identity data of an individual over its life

---

[5] http://www.biometrics.gov/Documents/IdMReport_22SEP08_Final.pdf

cycle. Many of the processes and technologies used to manage a person's identity may also be applied to NPEs.

Today, many system application owners and program managers create a digital representation of an identity by establishing and setting access privileges to enable application-specific processes. As a result, maintenance and protection of the identity is treated as secondary to the mission associated with the application. Unlike accounts used to log on to networks, systems, or applications, enterprise identity records are not tied to job title, job duties, location, or whether access is needed to a specific system. Those things may become attributes tied to an enterprise identity record, and may also become part of what uniquely identifies an individual in a specific application. Access control decisions will be based on the context and relevant attributes of a user, not solely the user's identity. The concept of an enterprise identity is that individuals will have a digital representation of themselves that can be leveraged across departments and agencies for multiple purposes, including access control.

A digital identity typically comprises a set of attributes that, when aggregated, uniquely identify a user within a system or enterprise. To establish trust in the individual represented by a digital identity, an agency may also conduct a background investigation. Attributes about an individual may be stored in various authoritative sources within an agency and linked to form an enterprise view of the digital identity. This digital identity may then be granted physical and logical access to applications and removed when access is no longer required.

With the establishment of an enterprise identity, it is important that policies and processes be developed to manage the life cycle of each identity. Management of an identity includes—

- The framework and scheme for establishing a unique digital identity,
- The ways identity data will be used,
- The protection of personally identifiable information,
- Controlling access to identity data,
- The policies and processes for management of identity data,
- Developing a process for remediation (i.e., solving issues or defects),
- Sharing authoritative identity data with applications that leverage it, and
- Revoking an enterprise identity.

As part of the framework for establishing a digital identity, diligence should be employed to limit data stored in each system to a set of attributes required to define the unique digital identity and still meet the requirements of integrated systems. A balance is needed between information stored or made available to internal and external systems, and the privacy of individuals.

Credential Management

According to NIST SP 800-63, a credential is an object that authoritatively binds an identity (and optionally, additional attributes) to a token possessed and controlled by a person.[6] Credential management supports the life cycle of the credential itself. In the federal government, examples of credentials include smart cards, private/public cryptographic keys, and digital certificates. The policies around credential management, from identity proofing to issuance to revocation, are fairly mature compared to the other parts of ICAM. Personal Identity Verification (PIV) standards are found in Federal Information Processing Standards (FIPS) Publication 201-1 and NIST SP 800-73-3 (hyperlinks found in appendix A). Federal Public Key Infrastructure (PKI) Common Policy and DOD Common Access Cards are examples of documents that are important for agency-specific credential implementations. Today, approximately 5 million PIV cards have been issued to federal employees and contractors (see figure 2).

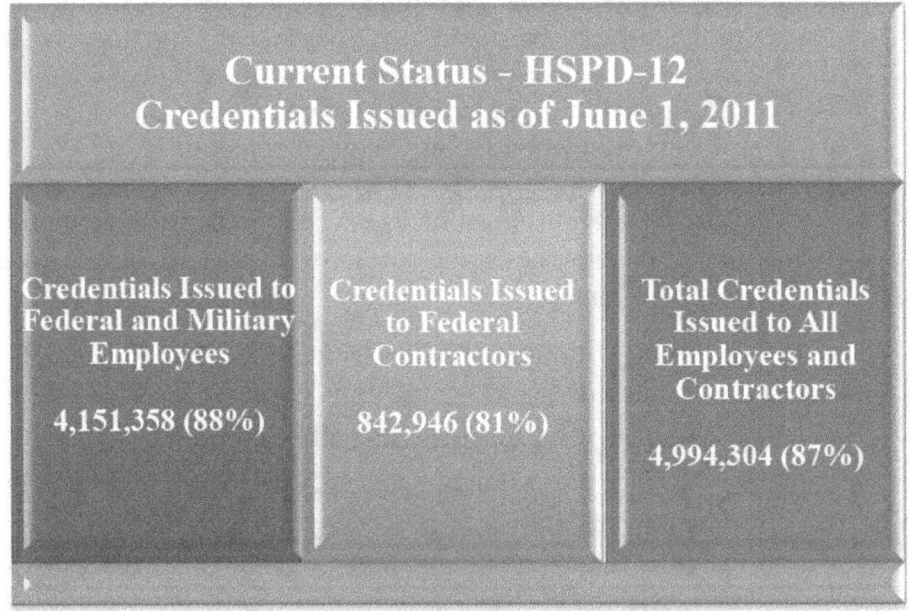

Current Status - HSPD-12
Credentials Issued as of June 1, 2011

Credentials Issued to Federal and Military Employees	Credentials Issued to Federal Contractors	Total Credentials Issued to All Employees and Contractors
4,151,358 (88%)	842,946 (81%)	4,994,304 (87%)

Figure 2: PIV cards data compiled from IDManagement.gov. Agency-specific status may be located at http://www.whitehouse.gov/omb/e-gov/hspd12_reports/. The percentages represent the percentage of each category obtaining credentials.

---

[6] The credentialing process principles and elements can also be applied for NPE digital identities; however, steps may vary during the credential issuance process (e.g., sponsorship, adjudication) based on an organization's security requirements.

Credentialing consists of an authorized employee sponsoring an individual or entity and justifying the need for the credential. Next, the individual enrolls for the credential, a process that typically consists of identity proofing and the capture of biographic and biometric data. The types of data required may depend on the credential type and the usage scenario. This step may be automatically completed based on data collected and maintained through identity management processes and systems, since enrollment for a credential requires much of the same data collection that is required as part of identity management. Subsequently, a credential will be produced and issued to the individual or NPE. As in the case of enrollment, these processes will vary based upon the credential type in question. Identity proofing, production, and issuance requirements for other credential types typically include a subset of the processes or technologies but follow the same general principles. Finally, a credential must be maintained over its life cycle, which might include revocation, reissuance/replacement, reenrollment, expiration, personal identification number reset, suspension, or reinstatement.

A key distinction in the life cycle management of credentials versus identities is that credentials expire. The attributes that form one's digital identity may change over time, but one's identity does not become invalid or terminated from a system perspective. Credentials, however, are usually valid for a predefined period, typically for five years. An example is certificates issued to an individual that expire based on the issuer's PKI Common Policy. While the identity of an individual does not change, the certificates associated with that individual can be revoked and new ones issued. This does not have a bearing on the individual's identity, as credentials are a tool that provides varying levels of assurance about the authentication of an individual.

Another key aspect of credential management is the security and protection of credentials, from issuance to termination. The trust in a credential depends on a multilayered approach to security that protects the credential as well as who can use the credential from attack. ICAM hinges on the level of trust in a credential and the uniformity of security and integrity across the security architecture in order to retain that trust throughout the use of the credential.

When the Working Group surveyed the IG community regarding the use of PIV card capabilities, the responses demonstrate OIGs are (1) currently using the PIV card for physical and/or logical access, (2) planning to use the PIV card for physical and/or logical access, (3) awaiting their parent agencies' direction on PIV use, and (4) currently have no plans to implement PIV card technology (see figure 3).

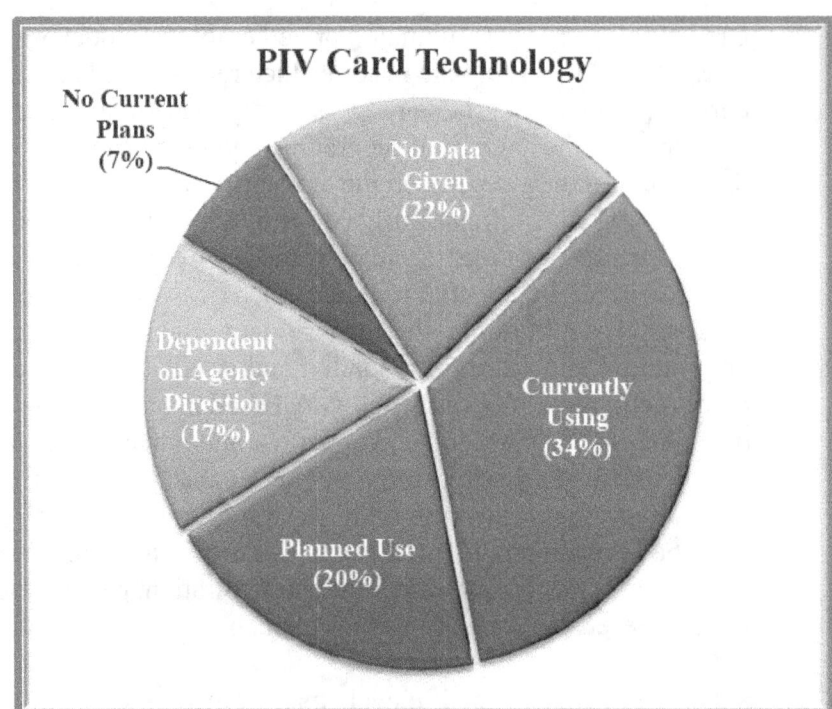

PIV Card Technology

No Current Plans (7%)

No Data Given (22%)

Dependent on Agency Direction (17%)

Currently Using (34%)

Planned Use (20%)

Figure 3: Results from the Cybersecurity Working Group survey. See appendix B for a summary of survey results.

Access Management

Access management is the management and control of the ways in which entities are granted access to resources. The purpose of access management is to ensure that the proper identity verification is made when an individual attempts to access security-sensitive buildings, computer systems, or data. It has two areas of operations: logical and physical access. Logical access is the access to an IT network, system, service, or application. Physical access is the access to a physical location such as a building, parking lot, garage, or office. Access management leverages identities, credentials, and privileges to determine access to resources by authenticating credentials. After authentication, a decision as to whether a person is authorized to access the resource can be made. These processes allow agencies to obtain a level of assurance in the identity of the individual by—

- Ensuring that all individuals attempting access are properly validated (authentication),

- Ensuring that all access to information is authorized (confidentiality),

- Protecting information from unauthorized creation, modification, or deletion (integrity),

- Ensuring that authorized parties are able to access needed information (reliability, maintainability, and availability), and

- Ensuring the accountability of parties when gaining access and performing actions (nonrepudiation).

In addition, access control sets the stage for activities outside of the traditional access control paradigm. One corollary to access management is the ability to ensure that all individuals attempting access have a genuine need. This is tied to authentication and authorization, but also to the business rules surrounding the data themselves. Privacy is provided by ensuring confidentiality and by refraining from collecting more information than necessary.

A key aspect of access management is the ability to leverage an enterprise identity for entitlements, privileges, multifactor authentication, roles, attributes, and different levels of trust. Logical and physical access are often viewed as the most significant parts of ICAM from a return-on-investment perspective. To maximize that return, a successful access management solution is dependent on identity, credentials, and attributes for making informed access control decisions, preferably through automated mechanisms. Without an access management solution, the vision and value of an identity access management initiative are diminished.

Recommendations

The Federal CIO Council, Information Security and Identity Management Committee, highlights some high-level considerations in its Federal Identity, Credential, and Access Management Roadmap and Implementation Guidance, Version 1.0. We recommend that OIGs consider the following guidance and practices, when applicable:

Recommendation #3: Refer to IDManagement.gov, a one-stop shop for citizens, businesses, and government entities interested in identity management activities, including topics related to HSPD-12; Federal PKI; identity, credential, and access management; and acquisitions.

Recommendation #4: Identify an application to employ two-factor identification to protect information based on NIST SP 800-63 guidance.

Recommendation #5: Evaluate personnel processes for hiring and separating employees.

Incident Detection and Handling

The IG community needs effective computer incident prevention, detection, and handling capabilities. Understanding that not all IT incidents can be prevented is critical to understanding the threats that face our networks today. An incident detection, reporting, and response capability is therefore necessary for effective network security.

Reports of security incidents from federal agencies are on the rise, increasing by more than 650% over the past 5 years.[7] The growing threats and increasing number of reported incidents highlight the need for a robust system of countermeasures to prevent incidents from occurring and to quickly detect and respond to incidents that cannot be prevented. However, serious and widespread information security control deficiencies continue to place federal assets at risk of misuse, sensitive information at risk of inappropriate disclosure, and critical operations at risk of disruption. Therefore, it is imperative that federal agencies implement an incident prevention, detection, and response program to ensure business continuity. Incident prevention includes, but is not limited to, boundary defenses, asset inventories, configuration management, user account management, and automated monitoring to provide real-time security status reporting. Incident detection and response covers automated detection, analysis, containment, eradication, and recovery.

IT security incidents, whether deliberate or unintentional, threaten the confidentiality, integrity, or availability of information and information resources. When an information security-related incident is suspected or discovered, personnel must immediately take steps to protect the information resource(s) at risk. Agencies must develop an incident response capability that enables coordinated efforts of a defined incident response team to respond to incidents. When an incident has been identified, the incident response team must have the knowledge and skills to follow standard procedures.

---

[7] *Cyber Security: Continued Attention Needed to Protect Our Nation's Critical Infrastructure and Federal Information Systems* (GAO-11-463T), March 16, 2011.

Incident Notification, Reporting, and Immediate Responses: A Government "Standard"

Many branches of the U.S. government, including the DOD, DHS, and the Intelligence Community, have drafted guidelines they believe should be used for effective network security. All the publications currently available from these entities are invaluable resources for designing and establishing network security plans and incident related procedures. Among agencies, there are still many differences in what the "standard" is for responding to incidents and how business should be conducted.

NIST has been developing a generalized standard for computer security incident handling that is applicable to any agency's architecture or network environment: NIST SP 800-61, Revision 1, *Computer Security Incident Handling Guide.*

NIST SP 800-61 provides guidelines for IT incident handling, particularly for analyzing incident-related data and determining the appropriate response to each incident. Because effective incident response is a complex undertaking, establishing a successful incident response capability requires substantial planning and resources. Continually monitoring threats through intrusion detection systems, full-time network packet capture, and other mechanisms is essential.[8]

As NIST SP 800-61 evolves into a potential Government Standard for incident response, it will need to keep up with the threats and trends. It is an effective starting point for the establishment of network incident response guidelines when combined with creating partnerships with agencies that have well-established programs in order to learn from their mistakes and grow from their innovation.

Establishing clear procedures for assessing the current and potential business impact of incidents is critical, as is implementing effective methods of collecting, analyzing, and reporting data. Each agency will need to understand the specific threats it faces, its critical assets and data, and to develop the appropriate tools for handling incidents. In the past, protection of agency data and network security was focused on quick remediation and reliance on antivirus and firewall technology. In the current cyber landscape, it can be impossible to get ahead of the threat, and active network investigations, network traffic control, and monitoring are often the only defense.

---

[8] Packet capture uses a computer program or a piece of computer hardware that can intercept and log traffic passing over a digital network or part of a network. As data streams flow across the network, the sniffer captures each packet and, if needed, decodes the packet's raw data, showing the values of various fields in the packet, and analyzes its content.

Incident Prevention

Securing Network Boundaries: The First Line of Defense Against Incidents

Most federal agencies have Internet-accessible computers on their networks in order to communicate with external business parties and with the public. These computers are prime targets for exploitation and thus are highly sought after by hackers. For example, organized crime groups and nation-states continuously scan the Internet for publicly accessible computers on federal agency networks. After finding and then exploiting a vulnerability on a publicly accessible computer, the cybercriminals use the exploited computer as a means to penetrate deeper into the agency's network to steal sensitive data or disrupt operations.

Thus, to prevent incidents resulting from unauthorized access, agencies need a system of defenses to both control the flow of traffic through their network borders and inspect its content. These boundary defenses must be multilayered—relying on, for example, firewalls, proxy servers, and network-based intrusion detection systems[9] [10]—to prevent or immediately detect intrusions into the agency's computer networks.

Configuration Management: Ensuring That Network Devices Are Securely Configured

CM is the process of establishing and controlling changes made to hardware and software throughout the life cycle of an information system. Often, computer operating systems are configured by the vendor for ease of deployment and ease of use rather than for security, leaving them exploitable in their default state. Hackers are aware of this industry practice and use automated attack programs to continuously scan federal agency networks for systems with vendor-configured (vulnerable) operating systems, which they can immediately exploit.

To reduce the number of incidents that result from exploiting this condition, the CIS has published recommended configuration settings, called benchmarks, for securing a wide variety of computer operating systems and other devices such as firewalls and routers.[11] A growing

---

[9] A firewall is a set of IT resources that separate and protect computer systems and data on an organization's internal networks from unauthorized access from an external network, such as the Internet.

[10] A proxy server is a computer system that acts as an intermediary for requests from local computers seeking resources or services from untrusted sources, such as from computers on the Internet. Key security features provided by proxy servers include filtering for malicious content and denying access to websites that are known sources of malware.

[11] According to CIS, its benchmarks are consensus-based, best practice security configuration guides both developed and accepted by government, business, industry, and academia. (See www.cisecurity.org for more information.) However, reference to CIS, a private organization, is made for informational purposes

number of federal agencies, including the National Aeronautics and Space Administration, have adopted the benchmarks as best practices for the secure configuration of computer operating systems and other network-attached devices.

In addition to CIS benchmarks, NIST through its National Checklist Program has defined a repository of vendor-developed checklists (benchmarks) for the secure configuration of computer operating systems and other network-attached devices. Moreover, the program has addressed the need for automating IT security processes through the SCAP to enable security tools to automatically check configuration by using the checklists.

Controlling Data Access Through Effective User Account Management

Sensitive data occur widely throughout federal computer systems and networks and include personally identifiable information, information controlled by International Traffic in Arms Regulations, and Export Administration Regulations, as well as third-party intellectual property. Accordingly, agencies must implement effective safeguards to prevent the loss or theft of these sensitive data.

For example, without proper safeguards such as restricting administrator or super-user account privileges and ensuring that only authorized personnel have system access, sensitive information, including law enforcement reports and personally identifiable information, could be disclosed for purposes of espionage, identity theft, or other types of criminal activity.[12] According to the System Administration Networking and Security (SANS) Institute,[13] a widely recognized authoritative source for best practices in IT security, the misuse of administrator privileges is the method most widely used by attackers to steal sensitive data from federal agencies. This problem is exacerbated when many users unnecessarily have administrative privileges. In such an environment, each account becomes a potential target for an attacker. Once an administrator account is compromised, the attacker has full access to the victim's machine, or to many machines when the attack involves accounts with domain administration privileges.

---

only and does not constitute an endorsement by CIGIE or any federal agency. Moreover, it does not imply that its recommendations are necessarily the most appropriate or best available.

[12] The super-user, unlike normal user accounts, can operate without limits, and misuse of the super-user account may result in spectacular disasters. User accounts are unable to destroy the system by mistake, so it is generally best to use normal user accounts whenever possible, unless you especially need the extra privilege.

[13] Reference to the SANS Institute, a private organization, is made for informational purposes only and does not constitute an endorsement by CIGIE or any federal agency. Moreover, it does not imply that its recommendations are necessarily the most appropriate or best available.

A second common way attackers gain unauthorized system access is by exploiting legitimate but inactive user accounts. This can occur when employees separate from an agency but their user accounts remain active. To prevent security incidents related to ineffective account management, it is necessary to (1) limit employee access to system rights and permissions employees need to perform their official duties, and (2) immediately deactivate all user accounts when employees separate from an agency.

<u>Knowing What Is on the Network: The Need for Inventories of Networked Devices</u>

An accurate and up-to-date inventory of an agency's network-attached devices, controlled by active monitoring and configuration management, can reduce the chance of attackers finding unauthorized and unprotected systems to exploit. For example, one common attack exploits the condition when new hardware is installed on a network one evening and not configured and patched with appropriate security updates until the following day. Attackers from anywhere in the world may quickly find and exploit such systems that are Internet-accessible. Furthermore, even in internal network systems, attackers who have already gained access may hunt for and compromise additional improperly secured systems. Some attackers use the local nighttime window to install backdoors on systems before they are hardened.[14]

Attackers also frequently look for experimental or test systems that are intermittently connected to the network but not included in an organization's standard asset inventory. Such experimental systems tend not to have as thorough security hardening or defensive measures as other systems on the network. Although these test systems do not typically hold sensitive data, they offer an attacker an avenue into the organization and a launching point for deeper network penetration.

According to the respondents to the Working Group's survey, the majority of the IG community maintains a complete and accurate list of hardware and software applications supporting IG community programs and operations (see figure 4).

---

[14] A backdoor in a computer system is a method of bypassing normal authentication, securing remote access to a computer, obtaining access to plaintext, and so on, while attempting to remain undetected. The backdoor may take the form of an installed program.

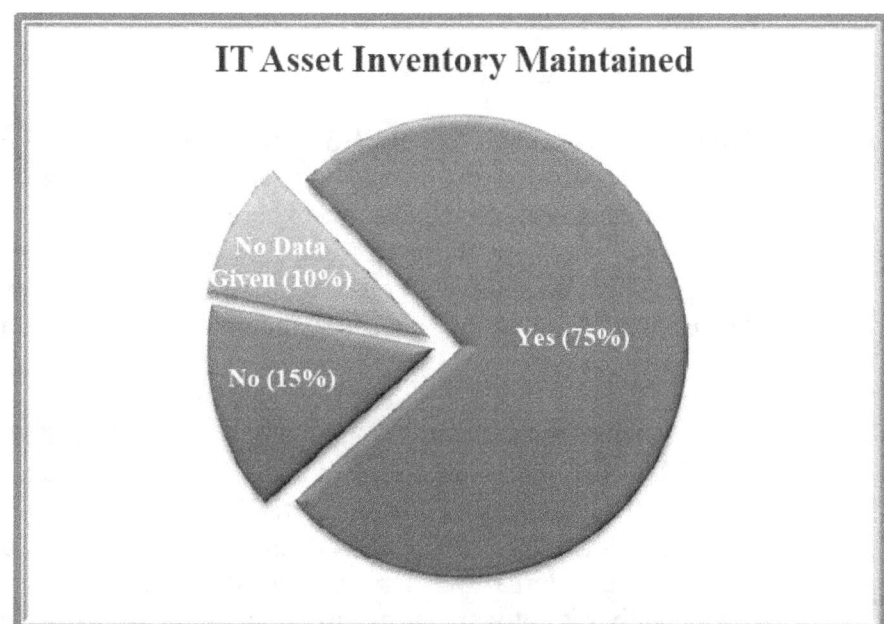

Figure 4: Results from the Cybersecurity Working Group survey.

Automating the Continuous Monitoring Program

Ensuring that federal information systems are adequately protected against ever-increasing threats requires mechanisms to establish and then continuously monitor (audit) key security controls. The goal of continuous monitoring is to determine whether a system's key IT security controls continue to be effective over time in light of changes to hardware or software. A well-designed and well-managed continuous monitoring program can transform an otherwise static security control assessment and risk determination process into a dynamic process that provides essential information about a system's security status on a real-time basis. This, in turn, enables officials to take timely risk mitigation actions and make risk-based decisions regarding the operation of the information system.

Automating the control monitoring process is essential because of the size, complexity, volatility, and interconnected nature of federal information systems. The SANS Institute has identified 20 critical IT security controls organizations should implement for effective cyber defense.[15] The SANS Institute recommends that federal agencies examine all 20 control areas against the current agency status and develop an agency-specific plan to implement the controls as a key component of an overall IT security program.

---

[15] "20 Critical Controls for Effective Cyber Defense: Consensus Audit Guidelines," SANS Institute, November 2009. Reference to the SANS Institute, a private organization, is made for informational purposes only and does not constitute an endorsement by CIGIE or any federal agency. Moreover, it does not imply that its recommendations are necessarily the most appropriate or best available.

Incident Detection

Incidents can occur from a myriad of complex sources and causes. Therefore, personnel monitoring an agency's IT infrastructure must have sufficient technical knowledge and experience to identify and analyze events and other incident-related data. Types of cybersecurity-related incidents include malicious code such as viruses, worms, and Trojan horses; denial-of-service attacks; unauthorized access; and inappropriate usage.[16] [17]

Indications of an incident can occur at different sources and levels. Some examples of incident indications include port-scanning activities reported by the intrusion detection system/intrusion prevention system; multiple or persistent failed login attempts from an unfamiliar system; unusual activity at an external web server; unusual deviation in network traffic flows; antivirus software alerts; user complaints of slow access or response; filenames with unusual characters; configuration changes in audit log files; and unusual numbers of bounced e-mails with suspicious content.

Information about zero-day exploits and known threats and vulnerabilities can also be a source for incident detection.[18] The possibility of incidents attributed to the "insider threat" should also be considered. Unauthorized access by insiders should prompt stronger policies concerning background investigations for personnel and stronger security controls on internal networks.

Incident Handling

According to OMB Memorandum 07-16, dated May 22, 2007, entitled *Safeguarding Against and Responding to the Breach of Personally Identifiable Information*, "when faced with a security incident, an agency must be able to respond in a manner protecting both its own information and helping to protect the information of others who might be affected by the incident. To address this need, agencies must establish formal incident

---

[16] Viruses, Worms, and Trojan Horses are all malicious programs that are purposely written to cause damage to a computer and/or information on the computer. They are also capable of slowing down the Internet, and they can use an individual's computer to spread themselves to friends, family, coworkers, or others.

[17] A denial-of-service attack is an attempt to make a computer resource unavailable to its intended users. Although the means to carry out, motives for, and targets of a denial-of-service attack may vary, it generally consists of the concerted efforts of a person, or multiple people to prevent an Internet site or service from functioning efficiently or at all, temporarily or indefinitely.

[18] A zero-day exploit is one that takes advantage of a security vulnerability on the same day the vulnerability becomes generally known. There are zero days between the time the vulnerability is discovered and the first attack.

response mechanisms. To be fully effective, incident handling and response must also include sharing information concerning common vulnerabilities and threats with those operating other systems and in other agencies. In addition to training employees on how to prevent incidents, all employees must also be instructed in their roles and responsibilities regarding responding to incidents should they occur." Possible metrics for incident-related data could include the number of incidents handled, time per incident, and assessments of each incident.

Best Practices for Incident Response, Network Defense, and Remediation

According to NIST SP 800-61, effective incident response has four phases: preparation; detection and analysis; containment, eradication, and recovery; and post-incident activity (see figure 5).

Figure 5: The Incident Response Life Cycle. *Source:* NIST SP 800-61, p. 3-1.

The initial phase involves establishing and training an incident response team, and acquiring the necessary tools and resources. During preparation, the organization also attempts to limit the number of incidents by selecting and implementing a set of controls based on the results of risk assessments. However, residual risk will inevitably persist after controls are implemented; furthermore, no control is foolproof. Detection of security breaches is thus necessary to alert the organization whenever incidents occur. In keeping with the severity of the incident, the organization can act to mitigate the impact of the incident by containing it and ultimately recovering from it. After the incident is handled, the organization issues a report that details the cause and cost of the incident and the steps the organization should take to prevent future incidents.[19]

Security Breach Notification

OMB Memorandum 07-16 requires agencies to develop and implement a breach notification policy. The term "personally identifiable information" refers to information that can be used to distinguish or trace an individual's identity. It includes information such as name, Social Security number, and biometric records. This information can be used alone or combined with other personal or identifying information that may

---

[19] NIST SP 800-61, *Computer Security Incident Handling Guide.*

be linked or linkable to a specific individual, such as date and place of birth and mother's maiden name. Agencies must report incidents involving personally identifiable information to the United States Computer Emergency Readiness Team.

External Notification of a Security Breach

Each agency should develop a breach notification policy and plan comprising the six elements discussed in OMB Memorandum 07-16 and listed below—

- Whether breach notification is required

- Timeliness of the notification

- Source of the notification

- Contents of the notification

- Means of providing the notification

- Who receives notification (public outreach v. internal communications)

When implementing the policy and plan, the agency head will make final decisions regarding breach notification. To ensure adequate coverage and implementation of the plan, each agency should establish an agency response team that includes the program manager of the program experiencing the breach; the CIO, Chief Privacy Officer or Senior Official for Privacy; Communications Office; Legislative Affairs Office; General Counsel; and the Management Office, which includes budget and procurement functions.[20]

Business Impact, Damage Assessment, and Lessons Learned

After the incident is handled, the agency should prepare a business impact and damage assessment. This assessment describes the cause and cost of the incident and the required steps to prevent future incidents. Items to consider when calculating the cost include damage to the agency's reputation; lost revenue; lost service and ability to operate; cost to remediate or replace information; cost to repair or replace damaged hardware; and potential fines, lawsuits, and legal fees. Conducting a "lessons learned" session with all involved personnel after an incident can help strengthen security measures while also improving the incident-handling process.

---

[20] "Safeguarding Against and Responding to the Breach of Personally Identifiable Information" (OMB Memorandum 07-16), May 22, 2007.

Results of CIGIE Cybersecurity Working Group Survey

To assess the IG community's incident detection and handling capability, we conducted a survey to identify common practices for identifying, containing, and responding to cybersecurity events. Survey results identified a number of areas where OIG security practices can be improved to enhance incident detection and handling capabilities. Specifically, survey results identified the following challenges—

- Thirty-one percent of respondents have not implemented incident detection and handling policies and procedures consistent with NIST SP 800-61, *Computer Security Incident Handling Guide*.

- Fifty-four percent of respondents do not periodically test their ability to identify, contain, and respond to cybersecurity events in accordance with local policy and procedures. Ineffective testing of incident detection and handling procedures could prevent OIGs from identifying and responding to system intrusion attempts in a timely manner.

- Forty percent of respondents have not implemented the capability to monitor their systems and networks for unauthorized access. Additionally, respondents do not have security event correlation capabilities to identify trends related to network intrusions or intrusion attempts.

- Twenty-seven percent of respondents do not periodically review system audit logs to identify unauthorized access attempts. Such reviews are critical for determining individual accountability, reconstructing security events, and identifying system performance issues.

- Seventeen percent of respondents have not implemented encryption controls to protect sensitive OIG data transmitted via email or to protect OIG data "at rest" from unauthorized access or disclosure.

- Seventeen percent of respondents do not have an accurate inventory of hardware, software, or applications supporting OIG critical operations. A complete list of hardware and software components is critical for protecting OIG systems in the event of a cybersecurity incident.

Recommendations

We recommend that OIGs consider implementing the following practices, when applicable:

Recommendation #6: Review CIS or other appropriate benchmarks for the secure configuration of critical network devices, including computer servers, firewalls, routers, and switches.

Recommendation #7: Monitor user account privileges for key OIG systems and limit privileged (e.g., administrator, superuser) system access to as few individuals as possible.

Recommendation #8: Implement a continuous security control monitoring program for key IT security controls, such as operating system configurations, system vulnerabilities, and software patch levels.

Scalable Trustworthy Systems

Due to differing IT models in operation, the IG community must be aware of the concepts of trust, scale, and composition when developing and implementing information systems. A clear understanding of these concepts is necessary so that organizations can maintain and improve the security posture of their IT environment when confronted with the adoption of emerging technologies, the demand for information sharing, and management of the technology refresh cycle.[21]

According to *A Roadmap for Cybersecurity Research*, released in 2009, trustworthiness is a multidimensional measure of the extent to which a system is likely to satisfy each of the following elements: system integrity, availability, survivability, data confidentiality, guaranteed real-time performance, accountability, attribution, and usability. Definitions of what trust means for each element and well-defined measures against which trustworthiness can be evaluated are fundamental to developing and operating trustworthy systems.[22]

As part of trustworthiness, IGs should identify Mission Critical (MC) applications and the infrastructure behind them. In the Continuity of Operations Plan (COOP) exercises that the DHS OIG participated in, all MC applications were identified but did not include all required underlying infrastructure to support those applications. IGs should

---

[21] The technology refresh cycle is the periodic replacement of IT and communications systems in response to changes in available technology.
[22] *A Roadmap for Cybersecurity Research*, Department of Homeland Security, November 2009.

recognize that it is critical to ensure that the disaster recovery infrastructure provides the same level of trust and the same security posture established for the MC applications during normal operations. This can be accomplished through architectural principles for the design and implementation of trustworthy, scalable systems. This makes emergency preparedness an extra requirement for trusted systems that can be addressed during the design phase. Designing systems for emergency upfront avoids the need to retrofit security measures after the disaster infrastructure has been already deployed. This also guarantees senior leadership the same level of risk that they are accustomed to during normal operations, removing concerns outside of the emergency at hand.

Scalability is the ability to satisfy given requirements as computer systems and networks expand in functionality, capacity, complexity, and scope of trustworthiness. Systems must be designed with scalability in mind because experience shows that scalability typically cannot be later retrofitted into a system. The primary concern of this area is scalability that preserves or enhances trustworthiness in real systems.

Composability is the ability to create systems and applications with predictably satisfactory behavior from components, subsystems, and other systems. To enhance scalability in complex, distributed applications that must be trustworthy, high-assurance systems[23] should be developed from a set of components and subsystems, each of which is itself suitably trustworthy, within a system architecture that inherently supports composition. Composition includes the ability to run software on different hardware, aided by virtualization, operating systems emulation, and portable code. [24] In addition, composition extends beyond the technical aspects of system design, and therefore, system requirements and system evaluations should compose accordingly. It is vital that new systems can be incrementally added, or composed, into a system of systems with some predictable confidence that the trustworthiness of the resulting systems of systems is not weakened.[25]

While members of the IG community may not be developing large-scale systems themselves, they will rely on services provided by other organizations specializing in these systems and networks. Examples

---

[23] High-assurance systems offer strong guarantees that the system conforms to specified requirements for confidentiality, integrity, availability, safety, reliability, maintainability, standards, documentation, procedures, and regulations.
[24] Virtualization is the use of virtual machines to let multiple network subscribers maintain individualized desktops on a single, centrally located computer or server. The central machine may be at a residence, business, or data center. Users may be geographically scattered but are all connected to the central machine by a proprietary local area network or wide area network or the Internet.
[25] *A Roadmap for Cybersecurity Research*, Department of Homeland Security, November 2009.

include mobile phone networks, cloud computing services,[26] agency intranets, and the Internet itself. While the IG community may not be responsible for the security of these systems, it is responsible for the security of the data it processes on the systems and must ensure that a sufficient level of trustworthiness is established.

The current framework developed to manage the risks to government information imposed by the IT gaps in composability and scalability is FISMA, supplemented by the NIST SP 800 series. However, it remains challenging to ensure the trustworthiness of systems based on whole-system evaluations imposed by FISMA, due to the lack of top-to-bottom and end-to-end analysis as well as the great burden on system administrators.

Approaches such as OMB's Trusted Internet Connection (TIC) and Federal Risk and Authorization Management Program (FedRAMP), from the CIO Council, attempt to alleviate the problem by providing high-assurance systems for Internet connectivity and cloud computing.[27]

The TIC initiative, headed by OMB and DHS, is a multifaceted plan for improving the federal government's security posture by reducing external connections, including those to the Internet. This consolidation will result in a common security solution that includes facilitating the reduction of external access points, establishing baseline security capabilities, and validating agency adherence to those security capabilities. Agencies participate in the TIC initiative either as TIC Access Providers (a limited number of agencies that operate their own capabilities) or by contracting with commercial managed trusted Internet protocol service providers through the GSA-managed Networx contract vehicle.[28] This effort addresses agencies' needs for connectivity by offering a trusted scalable architecture that enhances each individual agency's security posture.[29]

According to the survey respondents, there are variations in how OIGs connect to the Internet. Currently, most OIGs connect to the internet through their parent agency (see figure 6).

---

[26] Cloud computing services cover a wide range of scalable, on-demand infrastructure, service, and software solutions; it provides computation, software, data access, and storage services that do not require end-user knowledge of the physical location and configuration of the system that delivers the services.
[27] The Trusted Internet Connection initiative is meant to optimize individual external connections, including internet points currently in use by the Federal government of the United States.
[28] The GSA website says that the Networx program offers comprehensive, best value telecommunications providing for new technologies, industry partners, and ways to achieve a more efficient and effective government. Networx allows agencies to focus their resources on building seamless, secure operating environments while ensuring access to the best technology industry has to offer.
[29] A scalable architecture is the ability of a system, network, or process, to handle growing amounts of work in a graceful manner or its ability to be enlarged to accommodate that growth.

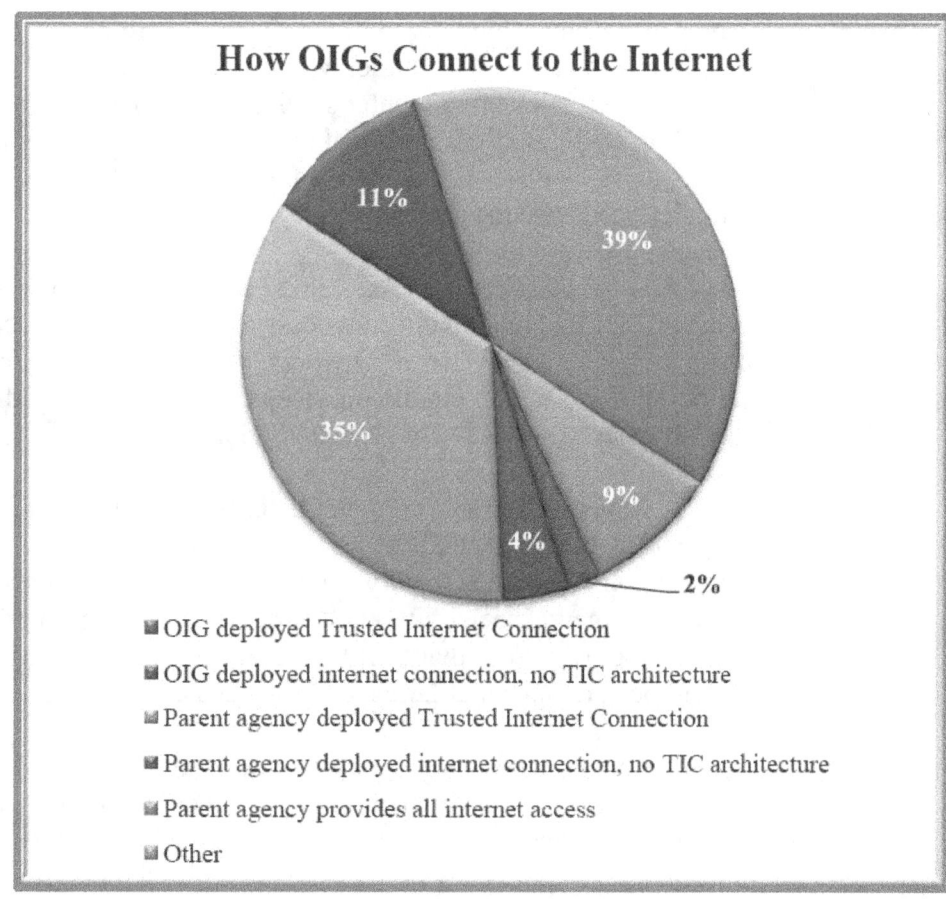

How OIGs Connect to the Internet

- OIG deployed Trusted Internet Connection
- OIG deployed internet connection, no TIC architecture
- Parent agency deployed Trusted Internet Connection
- Parent agency deployed internet connection, no TIC architecture
- Parent agency provides all internet access
- Other

Figure 6: Results from the Cybersecurity Working Group survey.

Beginning with the fiscal year 2012 budget, OMB requires agencies to consolidate their data centers and target cloud computing platforms as the primary operating model for new IT services.[30] With reduced IT budgets on agencies' immediate and long-term horizon, they must move to a new business model for delivering IT services. The IG community should focus on cloud computing as a primary option for new IT systems and services.

FedRAMP was established to provide a standard approach to assessing and authorizing cloud computing services and products. It allows joint authorizations and continuous security monitoring services for both government and commercial cloud computing systems intended for multiagency use. Joint authorization of cloud providers results in a common security risk model that can be leveraged across the federal government. This model provides a consistent baseline for cloud-based technologies, which ensures that their benefits are effectively integrated

---

[30] Cloud computing services cover a wide range of scalable, on-demand infrastructure, service, and software solutions; it provides computation, software, data access, and storage services that do not require end-user knowledge of the physical location and configuration of the system that delivers the services.

across the various cloud computing solutions currently proposed within the government. The risk model will also enable the government to quickly leverage cloud computer services following the "approve once and use often" method of ensuring that multiple agencies gain the benefit and insight of the FedRAMP's Authorization and Accreditation to the service provider's authorization packages.[31]

NIST SP 800-145 (Draft) defines cloud computing as a model for enabling convenient, on-demand network access to a shared pool of configurable computing resources, such as networks, servers, storage, applications, and services, which can be rapidly provisioned and released with minimal management or service-provider interaction.

Cloud computing systems are potentially beneficial for IG community members as they are scalable by design, offering the ability to distribute infrastructure resources rapidly and inexpensively. Cloud computing offers incremental scalability via "on-demand" allocation of computing and network resources that avoids typical system over-engineering and system performance that far exceeds its needs.

In addition, cloud computing may be beneficial in other aspects. For example, it would enable the government to contract out many IT computing services. Economies of scale could lower costs as having fewer but better trained people maintaining a few cloud systems should be significantly more efficient than maintaining many small, federal networks. Furthermore, having fewer systems, run by experts, using better hardware and software, may be significantly more secure.

Finally, cloud computing permits device independence through the use of virtualization technologies for servers, clients, and applications. Desktop services can be accessed from anywhere through web browsers and virtual remote desktop clients regardless of the client device type (e.g., laptops, smartphones, tablets). Consequently, cloud computing is capable of unifying multiple infrastructures under a common platform. Cloud computing services can be engineered for high reliability through the use of multiple, redundant zones, making the platform suitable for business continuity and disaster recovery. Cloud computing systems can be designed for high assurance without sacrificing their scalability and low cost.

Smaller OIGs may benefit most from cloud technologies that provide secure services that they cannot currently support internally. As reflected in figures 7 and 8, survey respondents stated that audit management software packages were fairly standardized within the IG community,

---

[31] http://www.cio.gov/pages.cfm/page/Federal-Risk-and-Authorization-Management-Program-FedRAMP

while a wide variety of software was used for case management. However, both audit management software and case management software could be used as a pilot for a multiagency-use cloud computing system.

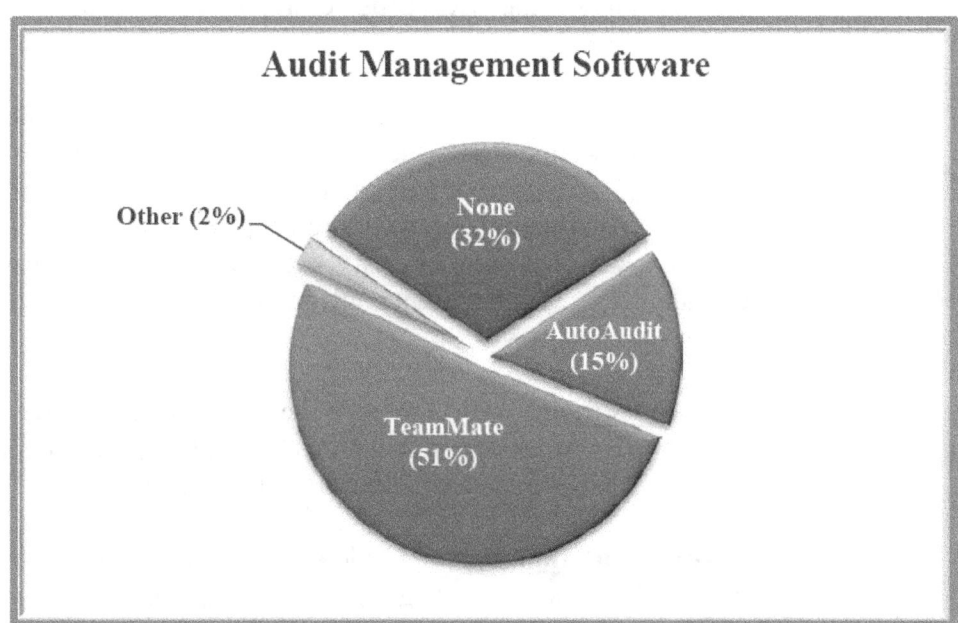

Figure 7: Results from the Cybersecurity Working Group survey.

Case Management Software	Number of Respondents	Percentage of Respondents
In-house Developed Application	11	26.82%
None	12	29.26%
AutoInvestigation	1	2.44%
Case Management System	1	2.44%
Case Management Tracking System	1	2.44%
CaseMap	1	2.44%
CMTS	1	2.44%
Concordance	1	2.44%
Law Enforcement Records System	1	2.44%
EDS	1	2.44%
Entellitrack	3	7.32%
I2MS	1	2.44%
IGCIRTS	1	2.44%
IG-Ideas	1	2.44%
Magnum	2	4.88%
Outsourced	1	2.44%
ProLaw	1	2.44%
Grand Total	**41**	**100.0%**

Figure 8: Results from the Cybersecurity Working Group survey.

Recommendations

We recommend that OIGs consider implementing the following practices, when applicable:

Recommendation #9: Carefully plan IT systems before deployment. New systems should enhance and maintain the security posture of the existing infrastructure and be capable of scaling according to projections. Requirements should include desired capabilities as well as nonfunctional requirements for system integrity, availability, survivability, data confidentiality, accountability, attribution, usability, and other critical needs.

Recommendation #10: Embrace the TIC architecture to enhance the security of network communications by ensuring that inbound and outbound data are properly monitored and secured.

Recommendation #11: Consider applications that could benefit from the FedRAMP cloud computing model. Finding common ground will improve efficiency of the OIG IT infrastructure and reduce its IT footprint and costs.

Reference List of Relevant Guidance, Laws, and Regulations

<u>**Presidential Directives and Executive Orders**</u>

Homeland Security Presidential Directive-12 (HSPD-12): Policy for a Common Identification Standard for Federal Employees and Contractors

Signed in 2004, HSPD-12 recognized that various forms of identification can be used to access secure facilities, which creates a potential risk for terrorist attack. HPSD-12 directed the government to eliminate those variations by creating a mandatory government standard for secure and reliable forms of identification issued by the government to employees and contractors. The policy is intended to enhance security, increase efficiency, reduce identify fraud, and protect personal privacy. It defined "secure and reliable forms of identification" as being issued based on sound criteria for verifying an individual's identity; strongly resistant to fraud and exploitation; rapidly authenticated electronically; and issued only by providers whose reliability was established with a specific process. NIST has published a variety of standards associated with HSPD-12 compliance.

The Comprehensive National Cybersecurity Initiative (CNCI) and The Cyberspace Policy Review (CPR)

- CNCI — In January 2008, President George W. Bush initiated the CNCI in National Security Presidential Directive 54/HSPD 23 to help secure the Nation in cyberspace. Major goals include establishing a front line of defense against immediate threats, defending against the full spectrum of threats, and strengthening the future cybersecurity environment. An unclassified summary describes the 12 initiatives established to achieve those goals. Those initiatives include managing the federal enterprise network as a single network enterprise with trusted Internet connections; deploying an intrusion detection system of sensors; connecting current cyber ops centers to enhance situational awareness; defining and developing enduring "leap-ahead" technology, strategies, and programs; defining and developing enduring deterrence strategies and programs; and defining the federal role for extending cybersecurity into critical infrastructure. http://www.whitehouse.gov/cybersecurity/comprehensive-national-cybersecurity-initiative

- CPR — In 2009, President Obama adopted recommendations set forth in the CPR. The CPR "outlines the beginning of the way forward towards a reliable, resilient, trustworthy digital infrastructure for the future." A few of the broad policies include leading from the top (e.g., appointing an executive branch Cybersecurity Coordinator); sharing responsibility for cybersecurity (e.g., federal government working closely with state and local governments and the private sector); creating effective information sharing and incident response; and encouraging innovation (e.g., establishing identity management mechanisms). The White House explains that the CNCI initiatives will evolve and support the achievement of many CPR recommendations. http://www.whitehouse.gov/assets/documents/Cyberspace_Policy_Review_final.pdf

Selected Office of Management and Budget Circulars and Memorandums

- **OMB Circular A-11**, *Preparation, Submission, and Execution of the Budget*. Part 2, Section 31.9, Management improvement initiatives and policies. Budget estimates should reflect efforts involving IT investments, E-government projects and strategy, commitment to privacy and reduction of improper payments, requirements of the *E-Government Act*, and a comprehensive understanding of OMB policies and NIST guidance. http://www.whitehouse.gov/sites/default/files/omb/assets/a11_current_year/s31.pdf

- **OMB Circular A-123**, *Management's Responsibility for Internal Control*. This circular provides guidance to federal managers on improving the accountability and effectiveness of federal programs and operations by establishing, assessing, correcting, and reporting on internal controls. In particular, Appendix III establishes that a minimum set of controls be included in information security programs. Moreover, each agency's program must implement policies and standards consistent with OMB, Department of Commerce, GSA, and Office of Personnel Management issuances. Agency heads are required to report annually on the effectiveness of the agency's security programs. http://www.whitehouse.gov/omb/circulars_a123_rev

- **OMB Circular A-130 Revised**, *Management of Federal Information Resources*. The U.S. Federal CIO Council's *Architecture Alignment and Assessment Guide* (2000) described OMB Circular A-130 as a "one-stop shopping document for OMB policy and guidance on information technology management." It establishes policies for the management of federal information resources government-wide, including the minimum controls to be included in federal automated information security programs and the assignment of federal agency responsibilities for the security of automated information. The circular also links agency automated information security programs and agency management control systems. http://www.whitehouse.gov/omb/circulars_a130_a130trans4/

- **OMB Memorandum 11-11**, *Continued Implementation of HSPD-12*. [32] This memorandum outlines a plan of action for agencies that will expedite the executive branch's full use of the credentials for access to federal facilities and information systems. http://www.whitehouse.gov/sites/default/files/omb/memoranda/2011/m11-11.pdf (See also OMB M-05-24.)

- **OMB Memorandum 11-02**, *Sharing Data While Protecting Privacy*. This memorandum directs agencies to find solutions that allow data sharing to move forward in a manner that complies with applicable privacy laws, regulations, and policies. http://www.whitehouse.gov/sites/default/files/omb/memoranda/2011/m11-02.pdf

---

[32] HSPD-12 applies to federal employees and contractors and requires (1) completion of background investigations, (2) issuance of standardized identity credentials, (3) use of the credentials for access to federal facilities, and (4) use of the credentials for access to federal information systems.

- **OMB Memorandum 10-27**, *Information Technology Investment Baseline Management Policy*. This memorandum provides policy direction regarding development of agency IT investment (both major and nonmajor investments) baseline management policies, and defines a common structure for IT investment baseline management policy with a goal of improving transparency, performance management, and effective investment oversight.
http://www.whitehouse.gov/sites/default/files/omb/assets/memoranda_2010/m10-27.pdf

- **OMB Memorandum 10-15**, *FY 2010 Reporting Instructions for the Federal Information Security Management Act and Agency Privacy Management*. This memorandum requires agencies to upload monthly inventory data feeds to CyberScope starting January 1, 2011. CyberScope is a web application developed by DHS in conjunction with the Department of Justice to handle manual and automated inputs of agency data for FISMA reporting.
http://www.whitehouse.gov/sites/default/files/omb/assets/memoranda_2010/m10-15.pdf

- **OMB Memorandum 08-22**, *Guidance on the Federal Desktop Core Configuration (FDCC)*. This guidance updates matters in **OMB Memorandum 07-11**, *Implementation of Commonly Accepted Security Configurations for Windows Operating Systems*, and discusses (1) Federal Desktop Core Configuration Major Version 1.0; (2) the SCAP validation requirement; (3) compliance, testing, and use of SCAP-validated tools for application providers supporting the federal government; (4) scope of "desktop" configuration; (5) revisions to part 39 of the Federal Acquisition Regulation; (6) the creation of the FDCC change control board; (7) updating FISMA guidance for FDCC; and (8) the policy utilization effort.
http://www.whitehouse.gov/sites/default/files/omb/memoranda/fy2008/m08-22.pdf

- **OMB Memorandum 08-05 and OMB Memorandum 08-27**

  o **Memorandum 08-05**, *Implementation of Trusted Internet Connections*. This memorandum announced the TIC initiative to optimize individual network services into a common solution for the federal government. The common solution facilitates the reduction of external connections, including Internet points of presence, to a target of 50. It required agencies to develop a comprehensive plan of action and milestones with a target completion date of June 2008.
  http://georgewbush-whitehouse.archives.gov/omb/memoranda/fy2008/m08-05.pdf

  o **Memorandum 08-27**, *Guidance for Trusted Internet Connection Compliance*. This memorandum instructs agencies identified as TIC Access Providers to ensure compliance with the TIC initiative, by (1) complying with critical TIC technical capabilities per the agencies' Statement of Capability; (2) continuing reduction and consolidation of external connections to identified TIC access points; (3) collaborating with the National Cyber Security Division; (4) executing a memorandum of agreement between DHS and the agency's CIO; and (5) executing a service-level agreement between DHS and the agency's CIO. http://www.whitehouse.gov/sites/default/files/omb/assets/omb/memoranda/fy200 8/m08-27.pdf

- **OMB Memorandum 07-18**, *Ensuring New Acquisitions Include Common Security Configurations*. This memorandum provides recommended language for agencies to use in solicitations to ensure that new acquisitions with Windows XP and Vista operating systems include configuration settings for FDCC settings discussed in M 07-11. http://www.whitehouse.gov/sites/default/files/omb/assets/omb/memoranda/fy2007/m 07-18.pdf

- **OMB Memorandum 07-16**, *Safeguarding Against and Responding to the Breach of Personally Identifiable Information*. This memorandum requires agencies to develop and implement a notification policy for internal and external breaches of personally identifiable information. It also requires agencies to develop policies concerning the responsibilities of individuals authorized to access personally identifiable information. http://www.whitehouse.gov/sites/default/files/omb/memoranda/fy2007/m07-16.pdf

- **OMB Memorandum 06-16**, *Protection of Sensitive Agency Information*. In addition to NIST's checklist for protection of remote information, this memorandum recommends that all departments and agencies take actions including (1) encrypting all data on mobile computers/devices that carry agency data unless the data are determined to be nonsensitive; (2) allowing remote access only with two-factor authentication where one of the factors is provided by a device separate from the computer gaining access; (3) using a "time-out" function for remote access and mobile devices, requiring user reauthentication after 30 minutes of inactivity; and (4) logging all computer-readable data extracts from databases holding sensitive information and verifying that each extract including sensitive data has been erased within 90 days or that its use is still required. http://www.whitehouse.gov/sites/default/files/omb/memoranda/fy2006/m06-16.pdf

- **OMB Memorandum 04-04**, *E-Authentication Guidance for Federal Agencies*. This memorandum requires agencies to review new and existing electronic transactions to ensure that authentication processes provide the appropriate level of assurance. It establishes and describes four levels of identity assurance for electronic transactions requiring authentication. Assurance levels also provide a basis for assessing credential service providers on behalf of federal agencies. The memorandum also assists agencies in determining their E-government authentication needs for users outside the executive branch. Further, it explains that agency business process owners bear the primary responsibility to identify assurance levels and strategies for providing them. The responsibilities set forth also extend to electronic authentication systems. http://www.whitehouse.gov/sites/default/files/omb/memoranda/fy04/m04-04.pdf

- **OMB Memorandum 03-22**, *OMB Guidance for Implementing the Privacy Provisions of the E-Government Act of 2002*. This memorandum directs agencies to conduct reviews of how information about individuals is handled within their agency when they use IT to collect new information, or when they develop or buy new IT systems to handle collections of personally identifiable information. http://www.whitehouse.gov/omb/memoranda_m03-22

- **OMB Memorandum 00-10**, *OMB Procedures and Guidance on Implementing the Government Paperwork Elimination Act (GPEA)*. This memorandum provides executive agencies with the guidance required under sections 1703 and 1705 of the GPEA. http://www.whitehouse.gov/omb/memoranda_m00-10/

Selected Federal Policies and Key Initiatives Affecting ICAM Implementation

The authorities and guidelines listed below,[33] as well as others discussed elsewhere in this report, reflect a small sample of relevant authorities related to ICAM implementation.

- *Privacy Act of 1974* (**5 U.S.C. § 552a**). The *Privacy Act,* in general, governs the collection, maintenance, use, and dissemination of personal information maintained by the federal government. In particular, the act covers systems of records that an agency maintains and retrieves by an individual's name or other personal identifier (e.g., Social Security number).

- *Health Insurance Portability and Accountability Act of 1996* (**P.L. 104-191**) (**HIPAA**). HIPAA protects the privacy of individually identifiable health information. The act also provides federal protections for personal health information held by covered entities and gives patients an array of rights with respect to that information.

- *Government Paperwork Elimination Act of 1998* (**P.L. 105-277**). GPEA requires federal agencies to allow individuals or entities that deal with the agencies the option to submit information or transact with the agency electronically, when practicable, and to maintain records electronically, when practicable. The act specifically states that electronic records and their related electronic signatures are not to be denied legal effect, validity, or enforceability merely because they are in electronic form and encourages federal government use of a range of electronic signature alternatives.

---

[33] List extracted from the Federal Chief Information Officers Council and the Federal Enterprise Architecture, *Federal Identity, Credential, and Access Management (FICAM) Roadmap and Implementation Guidance, Version 1.0*, section 2.3.3. (November 10, 2009). http://www.idmanagement.gov/documents/FICAM_Roadmap_Implementation_Guidance.pdf

- *Electronic Signatures in Global and National Commerce Act of 2000* **(P.L. 106-229).** This act was intended to facilitate the use of electronic records and signatures in interstate and foreign commerce by ensuring the validity and legal effect of contracts entered into electronically.

- **Executive Order 12977 - Interagency Security Committee.** This order established the Interagency Security Committee to develop standards, policies, and best practices for enhancing the quality and effectiveness of physical security in, and the protection of, nonmilitary federal facilities in the United States.

- **Executive Order 13467 - Reforming Processes Related to Suitability for Government Employment, Fitness for Contractor Employees, and Eligibility for Access to Classified National Security Information.** This order was established to ensure an efficient, practical, reciprocal, and aligned system for investigating and determining suitability for government employment, contractor employee fitness, and eligibility for access to classified information.

Selected National Institutes of Standards and Technology Publications

- **NIST SP 800-128**, *Guide for Security-Focused Configuration Management of Information Systems*. The publication includes guidelines for implementing CM security controls defined in NIST SP 800-53 and security controls related to managing the configuration of the system architecture and associated components for secure processing, storing, and transmitting of information. See the discussion on page 20 of this report for its applicability to configuration management practices. http://csrc.nist.gov/publications/nistpubs/800-128/sp800-128.pdf

- **NIST SP 800-73-3**, *Interfaces for Personal Identity Verification*. The guidance contains technical specifications to interface with PIV cards to retrieve and use identity credentials. The detailed publication comes in four parts: (1) End-Point PIV Card Application Namespace, Data Model and Representation; (2) PIV Card Application Card Command Interface; (3) PIV Client Application Programming Interface; and (4) The PIV Transitional Interfaces & Data Model Specification. http://csrc.nist.gov/publications/PubsByLR.html

- **NIST SP 800-61**, Revision 1, *Computer Security Incident Handling Guide*. This publication helps organizations mitigate the risks from computer security incidents and focuses on detecting, analyzing, prioritizing, and handling computer security incidents. http://csrc.nist.gov/publications/nistpubs/800-61-rev1/SP800-61rev1.pdf

- **NIST SP 800-53**, Revision 3, *Recommended Security Controls for Federal Information Systems and Organizations*. SP 800-53 includes a family of CM security controls. CM-8, Information System Component Inventory, requires organizations to develop, document, and maintain a current inventory of the components of an information system. http://csrc.nist.gov/publications/nistpubs/800-53-Rev3/sp800-53-rev3-final_updated-errata_05-01-2010.pdf

- **FIPS PUB 201-1**, *Personal Identity Verification (PIV) of Federal Employees and Contractors*. NIST published this Processing Standard to specify the architecture and technical requirement for a common identification standard for federal employees and contractors. http://csrc.nist.gov/publications/fips/fips201-1/FIPS-201-1-chng1.pdf. See also NIST SP 800-73.

Appendix B
Summary of Survey Results

The CIGIE Cybersecurity Working Group surveyed the IG community to solicit input about the current state of maintaining the integrity of OIG IT systems and carrying out its IT oversight responsibilities in the IG community. The working group invited 79 members of CIGIE to respond to the survey, which collected (1) demographics regarding personnel and budget and (2) information regarding various areas of IT. The results from the 41 survey respondents are summarized below.

General Demographics

1. How many total staff are employed by the OIG?

	Number of Respondents to Question	Percentage of Respondents to Question
1-10 people full-time	7	17.1%
11-50 people full-time	6	14.6%
51-99 people full-time	4	9.8%
100-249 people full-time	6	14.6%
250-500 people full time	9	22.0%
More than 500 people	6	14.6%
No Data Given	3	7.3%
Grand Total	**41**	**100.0%**

2. What is the annual OIG budget (including salary and benefits)?

	Number of Respondents to Question	Percentage of Respondents to Question
$250,001 to $500,000	1	2.4%
$500,001 to $999,999	2	4.9%
$1 million to $4,999,999	7	17.1%
$5 million to $9,999,999	9	22.0%
$10 million to $24,999,999	2	4.9%
$25 million to $49,999,999	6	14.6%
$50 million to $99,999,999	5	12.2%
Over $100 million	4	9.7%
No Data Given	5	12.2%
Grand Total	**41**	**100.0%**

3. What is the annual OIG IT budget (including salary and benefits)?

	Number of Respondents to Question	Percentage of Respondents to Question
Less than $100,000	10	24.4%
$100,000 to $250,000	1	2.4%
$250,001 to $499,999	4	9.8%
$500,000 to $999,999	3	7.3%
$1 million to $2,999,999	6	14.6%
$3 million to $9,999,999	10	24.4%
$10 million to $24,999,999	2	4.9%
No Data Given	5	12.2%
Grand Total	**41**	**100.0%**

4. Is the OIG IT budget allocated to sub-budgets?

	Number of Respondents to Question	Percentage of Respondents to Question
Yes	7	17.1%
No	30	73.2%
No Data Given	4	9.7%
Grand Total	**41**	**100.0%**

5. Does your OIG

	Number of Respondents to Question	Percentage of Respondents to Question
Have a fully staffed IT group which manages the day-to-day operations and is responsible for maintaining and supporting the OIG infrastructure?	12	29.3%
Rely on your parent agency to provide user and infrastructure support?	11	26.8%
Have a hybrid arrangement where the OIG and the parent agency share user and infrastructure support?	13	31.7%
Use a third-party to provide user and infrastructure support?	1	2.4%
No Data Given	4	9.8%
Grand Total	**41**	**100.0%**

6. If the OIG has its own IT group, how many full-time equivalents, excluding contractors, are, in any way, responsible for supporting or maintaining IT in the OIG? Please consider part-time staff in full-time equivalents.

	Number of Respondents to Question	Percentage of Respondents to Question
None	5	12.2%
One person less than full-time	2	4.9%
2 people	4	9.8%
3-5 people	4	9.8%
6-10 people	6	14.6%
11-20 people	6	14.6%
21-40 people	3	7.3%
41-80 people	1	2.4%
No Data Given	10	24.4%
Grand Total	**41**	**100.0%**

7. If the OIG has its own IT group, how many contractors does the OIG use to support or maintain the OIG's IT? Please consider part-time contractors in full-time equivalents.

	Number of Respondents to Question	Percentage of Respondents to Question
None	14	34.2%
One contractor	6	14.6%
2 to 5	7	17.1%
6 to 10	3	7.3%
More than 20 contractors	1	2.4%
No Data Given	10	24.4%
Grand Total	**41**	**100.0%**

Scalable Trustworthy Systems

Note: *For the next two questions, the number of responses is greater than the 41*
respondents because respondents could pick more than one choice.

1. What technology does the OIG use to support telework policy?

	Number of Respondents to Question	Percentage of Respondents to Question
Laptop	34	25.2%
Virtual Desktop (CITRIX or Remote Desktop)	17	12.6%
Web based services	18	13.3%
Removable storage (e.g., thumb drives)	23	17.0%
Physical Tokens	21	15.6%
Virtual Tokens	5	3.7%
Smart Cards	6	4.4%
Encrypted Storage on Device	1	0.7%
Mobile Device	2	1.5%
Remote Access Server	1	0.7%
Virtual Private Network	3	2.2%
No Data Given	4	3.1%
Grand Total	**135**	**100.0%**

2. What architecture and provider is used for Internet connectivity?

	Number of Respondents to Question	Percentage of Respondents to Question
OIG-deployed TIC	1	2.1%
OIG-deployed Internet connection, no TIC architecture	2	4.3%
Parent agency-deployed Trusted Internet Connection	16	34.1%
Parent agency-deployed Internet connection, no TIC architecture	5	10.6%
Parent agency provides all Internet access	19	40.4%
No Data Given	4	8.5%
Grand Total	**47**	**100.0%**

Identity, Credential, and Access Management

1. Has the OIG implemented a Federal ICAM Program Office or budget line item to support federal ICAM?

	Number of Respondents to Question	Percentage of Respondents to Question
Yes	10	24.4%
No	27	65.85%
No Data Given	4	9.75%
Grand Total	**41**	**100.0%**

2. Has the OIG supplemented OMB Memorandum 05-24, FIPS 201, and NIST guidance with its own policies, directives, or governance procedures to support federal ICAM?

	Number of Respondents to Question	Percentage of Respondents to Question
Yes	6	14.6%
No	30	73.2%
No Data Given	5	12.2%
Grand Total	**41**	**100.0%**

3. Does the OIG use PIV credentials for network/domain authentication?

	Number of Respondents to Question	Percentage of Respondents to Question
Yes	10	24.4%
No	27	65.8%
No Data Given	4	9.8%
Grand Total	**41**	**100.0%**

4. Has the OIG met all of the details in OMB directives for implementing HSPD-12 and federal ICAM?

	Number of Respondents to Question	Percentage of Respondents to Question
Yes	10	24.4%
No	23	56.1%
No Data Given	8	19.5%
Grand Total	**41**	**100.0%**

Incident Detection and Handling

1. Does the OIG have policies and procedures supporting an Incident Detection and Handling program for OIG systems?

	Number of Respondents to Question	Percentage of Respondents to Question
Yes	26	63.4%
No	11	26.8%
No Data Given	4	9.8%
Grand Total	**41**	**100.0%**

    a. If "Yes," are the policies and procedures consistent with NIST SP 800-61 Computer Security Incident Handling Guide and/or service-level agreements with the OIG's parent agency?

	Number of Respondents to Question	Percentage of Respondents to Question
Yes	25	61.0%
No Data Given	1	2.4%
Answer to previous question was "No" or not data given	15	36.6%
Grand Total	**41**	**100.0%**

2. Does the OIG periodically test its Incident Handling and Detection procedures to ensure they meet OIG security objectives?

	Number of Respondents to Question	Percentage of Respondents to Question
Yes	16	39.0%
No	20	48.8%
No Data Given	5	12.2%
Grand Total	**41**	**100.0%**

3. Who performs Incident Detection services for OIG systems?

	Number of Respondents to Question	Percentage of Respondents to Question
Parent Agency	17	41.5%
OIG	6	14.6%
Parent Agency/OIG	14	34.1%
No Data Given	4	9.8%
Grand Total	**41**	**100.0%**

4. Who performs Incident Handling services for OIG systems?

	Number of Respondents to Question	Percentage of Respondents to Question
Parent Agency	12	29.3%
OIG	5	12.2%
Parent Agency/OIG	20	48.8%
No Data Given	4	9.7%
Grand Total	**41**	**100.0%**

5. Does the parent organization keep the OIG informed of security incidents?

	Number of Respondents to Question	Percentage of Respondents to Question
Yes	36	87.8%
No	1	2.4%
No Data Given	4	9.8%
Grand Total	**41**	**100.0%**

5.a. If "No," provide potential recommendations for improving notification of security related events.

The one "No" respondent did not provide potential recommendations.

6. Does the OIG maintain a complete and accurate listing of hardware/software/applications supporting OIG programs and operations?

	Number of Respondents to Question	Percentage of Respondents to Question
Yes	31	75.6%
No	6	14.6%
No Data Given	4	9.8%
Grand Total	**41**	**100.0%**

7. What method does the OIG use for encryption of email messages?

	Number of Respondents to Question	Percentage of Respondents to Question
None	7	12.7%
PKI	16	29.1%
PGP	4	7.3%
Manual/attachment only encryption (i.e., Winzip or other)	16	29.1%
Other Encryption Method	7	12.7%
No Data Given	5	9.1%
Grand Total	**55**	**100.0%**

Note: For question 7, the number of responses is greater than the 41 respondents because respondents could pick more than one choice.

8. Does the OIG utilize encryption to protect OIG sensitive data from unauthorized access and disclosure?

	Number of Respondents to Question	Percentage of Respondents to Question
Yes	30	73.2%
No	5	12.2%
No Data Given	6	14.6%
Grand Total	**41**	**100.0%**

8.a. If "Yes," indicate whether the OIG has implemented these protections on end user computers and storage devices.

	Number of Respondents to Question	Percentage of Respondents to Question
Whole disk encryption has been implemented on end user computers and storage devices.	23	56.1%
Whole disk encryption has NOT been implemented on end user computers and storage devices.	5	12.2%
Answer to previous question was "Yes" but respondent did not answer 8.a	2	4.9%
Answer to previous question was "No" or no data given	11	26.8%
Grand Total	**41**	**100.0%**

9. Does the OIG have intrusion detection capability to monitor traffic on the OIG's internal network?

	Number of Respondents to Question	Percentage of Respondents to Question
Yes	22	53.7%
No	14	34.1%
No Data Given	5	12.2%
Grand Total	**41**	**100.0%**

10. Does the OIG have a security event correlation capability to identify security incidents?

	Number of Respondents to Question	Percentage of Respondents to Question
Yes	22	53.7%
No	14	34.1%
No Data Given	5	12.2%
Grand Total	**41**	**100.0%**

11. Does the OIG consistently review its system audit logs to detect unauthorized access attempts to OIG systems?

	Number of Respondents to Question	Percentage of Respondents to Question
Yes	26	63.4%
No	9	22.0%
No Data Given	6	14.6%
Grand Total	**41**	**100.0%**

Emergency Management

It is the policy of the United States to have in place a comprehensive and effective program to ensure continuity of essential Federal functions under all circumstances. As a baseline of preparedness for the full range of potential emergencies, all Federal agencies shall have in place a viable COOP capability which ensures the performance of their essential functions during any emergency or situation that may disrupt normal operations.

COOP planning is simply a "good business practice"—part of the fundamental mission of agencies as responsible and reliable public institutions. For years, COOP planning had been an individual agency responsibility primarily in response to emergencies within the confines of the organization. The content and structure of COOP plans, operational standards, and interagency coordination, if any, were left to the discretion of the agency.

The changing threat environment and recent emergencies, including localized acts of nature, accidents, technological emergencies, and military or terrorist attack-related incidents, have shifted awareness to the need for COOP capabilities that enable agencies to continue their essential functions across a broad spectrum of emergencies. Also, the potential for terrorist use of weapons of mass destruction has emphasized the need to provide the President a capability which ensures continuity of essential government functions across the Federal Executive Branch.

To provide a focal point to orchestrate this expanded effort, *Presidential Decision Directive 67* established the Federal Emergency Management Agency (FEMA) as the Executive Agent for Federal Executive Branch COOP. Inherent in that role is the responsibility to formulate guidance for agencies to use in developing viable, executable COOP plans; facilitate interagency coordination as appropriate; and oversee and assess the status of COOP capability across the Federal Executive Branch. Additionally, each agency is responsible for appointing a senior Federal government executive as an Emergency Coordinator to serve as program manager and agency point of contact for coordinating agency COOP activities.

Summary of OIG's Disaster Preparedness Activities

DHS OIG participated in the COOP Eagle Horizon 2011 Exercise (EH-11) on Thursday, June 23, 2011. EH-11 is the annual, integrated continuity exercise for all Federal Executive Branch departments and agencies, as mandated by the National Continuity Policy Implementation Plan, Federal Continuity Directive 1, and National Security Presidential Directive-51/Homeland Security Presidential Directive-20, *National Continuity Policy.*

The EH-11 exercise presents an opportunity to test organizational readiness and the capability to execute continuity plans and programs. During the exercise, the Office of Management, Information Technology Division (ITD), Infrastructure Branch was responsible for three distinct responsibilities. The first task was the transferring (failover) of IT Mission Essential Functions (MEFs)[34] from OIG's primary location at Headquarters in Washington, DC to the redundant disaster recovery site in Frisco, Texas. Second, once the continuity of IT services was ensured through the successful transition of operations to the redundant site, actions were started to prepare for the full recovery and reconstitution (failback) of IT MEFs back to HQ. The third task was to form a technical Advance Response Team to travel to the FEMA, National Emergency Training Center in Emmitsburg, Maryland to coordinate and support the OIG Emergency Relocation Group. The Advance Response Team was responsible for establishing communications and connectivity to the network as well as supporting technical issues from ERG members.

Since 2004, DHS ITD has been heavily involved in preparing and testing for COOPs in the OIG. In June 2009, ITD implemented its first major testing of the IT systems by failing over core network infrastructure and email services to the disaster data center. While the tests were viewed as a success, there were several components that did not behave as expected.

ITD began the planning for the 2011 COOP scenario in November of 2009. Once MEFs were identified by senior management, several design changes were required to be made to the existing underlying architecture for the new requirements as well as to correct the problems that were identified in the previous 2009 COOP tests. Starting in March of 2010, ITD began to implement these changes on a monthly basis on the last Friday of each month during the monthly maintenance. In April of 2011, the final changes were implemented on a bi-weekly basis.

OIG's Lesson's Learned

DHS OIG has fully participated in the annual DHS Eagle COOP exercise since 2005. As a major lead in COOP preparedness, ITD's goals included establishing a redundant data center for COOP and full failover of all mission critical applications and infrastructure to the designated location. During the last seven years, OIG has continually improved upon its disaster readiness through the continual improvements made to its COOP program. The following chart summarizes lessons learned during the continual planning for and execution of the annual COOP exercises.

---

[34] IT MEFs are categorized into three areas: 1) Core Network Infrastructure, 2) Messaging and Communications, and 3) Files and Information.

EVENT	LESSON LEARNED
During first COOP exercise, DHS OIG identified all MC applications but did not include the required underlying infrastructure to support the applications. For example, components such as active directory and SharePoint were initially overlooked.	Identify MC applications as well as the infrastructure behind them.
OIG users were not identified as MC and therefore were not available during COOP tests.	After identifying MC systems, ensure proper authoritative personnel are aware of the selected systems as well as those that are not selected systems.
Some essential account information, infrastructure diagrams and support information were not accessible with the primary infrastructure offline.	Identify support information from prior COOP exercises.
During first COOP exercise, applications were tested for failover prior to the actual test and were successful. When the test date came, several tests failed for various reasons.	Create specific test plans customized for each MC component and involve multiple personnel in the documentation, testing, and execution of each specific MC application; rotate a team to document and test the plan.
Large call volume and inundation due to basic requests.	Set realistic expectations for user population on resource availability; provide a detailed explanation to your user base specifically stating what resources will be available.
Several Virtual Private Network (VPN) users waited until the exercise date to test their VPN authentication and were unprepared to connect to the VPN.	Allow your user community that will be teleworking to connect, verify, and test their capability to the network at least a week prior to the test.
Could not calculate the increase or decrease in load on test date because the team did not know how many people connected to the VPN on a daily basis.	Establish baseline statistics and metrics of MC applications and infrastructure prior to the actual test so that you can capture statistics.

Appendix D
Contributors to this Report

The CIGIE Cybersecurity Working Group consisted of representatives of the following Offices of the Inspectors General:

Corporation for National and Community Service
Department of Agriculture
Department of Commerce
Department of Defense
Department of Education
Department of Health and Human Services
Department of Homeland Security
Department of State
Department of Transportation
Department of Veterans Affairs
Farm Credit Administration
Federal Deposit Insurance Corporation
National Aeronautics and Space Administration
National Security Agency
Securities Exchange Commission
Small Business Administration
Social Security Administration
Tennessee Valley Authority
Treasury Inspector General for Taxation Administration
United States International Trade Commission
United States Postal Service

www.ingramcontent.com/pod-product-compliance
Lightning Source LLC
Chambersburg PA
CBHW081616170526
45166CB00009B/2986